Farooq Ahmed Mehdi
Safa aldeen Al Naimi

Control system of fluidized catalytic cracking unit (FCCU)

Farooq Ahmed Mehdi
Safa aldeen Al Naimi

Control system of fluidized catalytic cracking unit (FCCU)

Multivariable control, Interaction and decoupling system in FCC unit

LAP LAMBERT Academic Publishing

Impressum/Imprint (nur für Deutschland/only for Germany)
Bibliografische Information der Deutschen Nationalbibliothek: Die Deutsche Nationalbibliothek verzeichnet diese Publikation in der Deutschen Nationalbibliografie; detaillierte bibliografische Daten sind im Internet über http://dnb.d-nb.de abrufbar.
Alle in diesem Buch genannten Marken und Produktnamen unterliegen warenzeichen-, marken- oder patentrechtlichem Schutz bzw. sind Warenzeichen oder eingetragene Warenzeichen der jeweiligen Inhaber. Die Wiedergabe von Marken, Produktnamen, Gebrauchsnamen, Handelsnamen, Warenbezeichnungen u.s.w. in diesem Werk berechtigt auch ohne besondere Kennzeichnung nicht zu der Annahme, dass solche Namen im Sinne der Warenzeichen- und Markenschutzgesetzgebung als frei zu betrachten wären und daher von jedermann benutzt werden dürften.

Coverbild: www.ingimage.com

Verlag: LAP LAMBERT Academic Publishing GmbH & Co. KG
Dudweiler Landstr. 99, 66123 Saarbrücken, Deutschland
Telefon +49 681 3720-310, Telefax +49 681 3720-3109
Email: info@lap-publishing.com

Approved by: Baghdad, University of Technology, Diss., 2007

Herstellung in Deutschland:
Schaltungsdienst Lange o.H.G., Berlin
Books on Demand GmbH, Norderstedt
Reha GmbH, Saarbrücken
Amazon Distribution GmbH, Leipzig
ISBN: 978-3-8383-9939-3

Imprint (only for USA, GB)
Bibliographic information published by the Deutsche Nationalbibliothek: The Deutsche Nationalbibliothek lists this publication in the Deutsche Nationalbibliografie; detailed bibliographic data are available in the Internet at http://dnb.d-nb.de.
Any brand names and product names mentioned in this book are subject to trademark, brand or patent protection and are trademarks or registered trademarks of their respective holders. The use of brand names, product names, common names, trade names, product descriptions etc. even without a particular marking in this works is in no way to be construed to mean that such names may be regarded as unrestricted in respect of trademark and brand protection legislation and could thus be used by anyone.

Cover image: www.ingimage.com

Publisher: LAP LAMBERT Academic Publishing GmbH & Co. KG
Dudweiler Landstr. 99, 66123 Saarbrücken, Germany
Phone +49 681 3720-310, Fax +49 681 3720-3109
Email: info@lap-publishing.com

Printed in the U.S.A.
Printed in the U.K. by (see last page)
ISBN: 978-3-8383-9939-3

Acknowledgment

I wish to express my sincere gratitude and thankfulness to
Prof. Dr. Neran A. Khalil
Department of Chemical Engineering -University of Technology.

Assistant Lecturer Shaymaa H. Khazaal
Department of Applied Science -University of Technology.

My deep thanks and appreciation to
Ms. Tatiana Costandachi
Acquisition Editor in LAMBERT Academic Publishing.

Dedication

To

All staff of Chemical Engineering Department - University of Technology, Baghdad -Iraq

Table of contents

List of tables

List of figures

Nomenclature

Greek Symbols

List of Abbreviations

List of tables

List of figures

Nomenclature

Symbol	Definition	Units
(A)	Fuzzy subset	[−]
B	Final steady state value of process reaction curve method	[−]
Cp_{rg}	Specific heat of regenerated catalyst	[kJ/kg.$^{\circ}$C]
Cp_f	Specific heat of fresh feed	[kJ/kg.$^{\circ}$C]
Cp_p	Specific heat of product	[kJ/kg.$^{\circ}$C]
Cp_s	Specific heat of spend catalyst	[kJ/kg.$^{\circ}$C]
$D_1(s)$	Dynamic element (Decoupler) for loop 1	[−]
$D_2(s)$	Dynamic element (Decoupler) for loop 2	[−]
E	Error	[$^{\circ}$C]
F_a	Mass flow rate of air	[kg /sec]
F_s	Mass flow rate of spend catalyst	[kg /sec]
F_{fg}	Mass flow rate of flue gases	[kg /sec]
F_f	Mass flow rate of fresh feed	[kg /sec]
F_p	Mass flow rate of product	[kg /sec]
F_{rg}	Mass flow rate of regenerated catalyst	[kg /sec]
$F_a(s)$	Transfer function of air	[kg /sec]
$F_{rg}(s)$	Transfer function of regenerated catalyst	[kg /sec]
$G_c(s)$	Transfer function of controller	[−]
$G_p(s)$	Transfer function of process	[−]
$H_{11}(s)$	Transfer functions between $T_{rc}(s)$ and $F_{rg}(s)$	[−]
$H_{12}(s)$	Transfer functions between $T_{rc}(s)$ and $F_a(s)$	[−]

$H_{21}(s)$	Transfer functions between $T_{rg}(s)$ and $F_{rg}(s)$	[−]
$H_{22}(s)$	Transfer functions between $T_{rg}(s)$ and $F_a(s)$	[−]
K_c	Proportional gain	[psig/$^{\circ}$C] or [−]
K	Steady state gain of the Process Reaction Curve method	[$^{\circ}$C/kg/sec]
M_s	Mass of spent catalyst	[kg]
M_{fg}	Mass of flue gases	[kg]
M_p	Mass of reactor product	[kg]
M_{rg}	Mass of regenerated catalyst	[kg]
s	Laplacian variable	[s^{-1}]
S	Slope of the tangent at the point of Inflection of the process reaction curve method	[$^{\circ}$C/sec]
T_a	Air temperature	[$^{\circ}$C] or [K]
T_{rc}	Reactor temperature	[$^{\circ}$C] or [K]
T_{rg}	Regenerator temperature	[$^{\circ}$C] or [K]
t	Time	[sec]
t_d	Time delay	[sec]
U_s	Scalar control unit	[−]
XG	Controller gain in Fuzzy logic control	[−]

Greek Symbols

Symbol	Definition	Units
ΔH_R	Heat of reaction	[kJ/kg]
ΔH_C	Heat of combustion	[kJ/kg]
Λ	Relative gain array	[−]
λ_{ij}	Elements of relative gain array	[−]
λ_{11}	Relative gain between T_{rc} and F_{rg}	[−]
λ_{12}	Relative gain between T_{rc} and F_a	[−]
λ_{21}	Relative gain between T_{rg} and F_{rg}	[−]
λ_{22}	Relative gain between T_{rg} and F_a	[−]
τ	Time constant	[sec]
τ_D	Derivative time	[sec]
τ_I	Integral time	[scc]
υ	Universe of discourse	[−]
μ	Membership function	[−]

List of Abbreviations

Symbol	Definition
FCCU	Fluidized Catalytic Cracking Unit
FLC	Fuzzy Logic Control
IMC	Internal Model Control
ISE	Integral of Square of Error
MATLAB	Matrix Laboratory
MIMO	Multi-input & Multi-output
NB	Negative Big
NCB	Negative change of Error Big
NEB	Negative Error Big
NES	Negative Error Small
NS	Negative Small
NUB	Negative control action Big
NUS	Negative control action Small
PB	Positive Big
PEB	Positive Error Big
PES	Positive Error Small
PID	Proportional-Integral-Derivative
PRC	Process Reaction Curve
PS	Positive Small
PCB	Positive change of Error Big
PCS	Positive change of Error Small

PUB	Positive control action Big
PUS	Positive control action Small
RGA	Relative Gain Array
ZC	Zero change of error
ZE	Zero Error
ZU	Zero control action

Chapter One: Introduction

Worldwide, the fluidized catalytic cracker unit (FCCU) is the principal part in the modern refineries. Its aim is to convert heavy hydrocarbon petroleum fractions into more usable products such as gasoline at high octane number, middle distillates, and light gases.

1.1 FCC Development History

The first FCC unit was built in USA in May 25, 1942 and its Location is in Baton Rouge refinery of Standard Oil of Louisiana. Many developments and researches have been published in FCC unit relating design, operation, feed specification, catalyst type Kinetics, dynamics and control system. These researches were participated actively in increasing the efficiency of gasoline production where the aim of the FCC unit. Now, worldwide the half of gasoline production that supplies the world requirements was produced by FCC unit.

Many companies were took part and played important role in FCC unit developments. Some of these companies focus on the design of reactor and regenerator or the catalyst used such as universal oil products (UOP) built a first FCC unit in 1947, designed side by side type that is mean, the reactor is beside the regenerator in middle of 1950 where it is still widely used in world and proved its efficiency in gasoline production. Kellogg Company designed FCC unit Model III in 1947 and introduced another design where the reactor at top of regenerator called orthoflow FCC unit in 1951. Shell Oil Company introduced riser cracking in 1956. Exxon Company designed FCC unit Model IV in 1953. The famous researchers for the Exxon precursor, Standard Oil are Donald L. Campbell, Homer Z. Martin, Egar V. Murphy, and Charles W. Tyson where their researches led to improve and raise the modern type of FCC unit.

In other side, development in used catalyst played an essential role in improvement the efficiency of FCC unit and gasoline production where in 1940, the catalyst that used was synthetic silica-alumina particles but discovering zeolite particles

1

in 1960 and combination with silica-alumina made revolution in catalytic creaking industry because the specification of zeolite made this type widely used in FCC unit.

1.2 Process Flow Description

The process flow sheet for FCC unit is shown in Figure (1.1). Fresh hydrocarbon feed (Atmospheric gas oil) is preheated then pumped to the riser at 200 °C (473 K) where injected into the riser bottom, converted into small particles and contacted with the high temperature regenerated catalyst from the regenerator at 720 °C (993 K) where the residence time is 5 seconds. Feed should not be heated above 397 °C (670 K) to prevent coking of the heating coils. Unlike other reactors, preheat is often just a control feature.

The heat from the catalyst vaporizes the feed and brings to the desired reaction temperature at 500 °C (773 K) then mixture of catalyst and hydrocarbons vapor passed to the reactor. The cracking reactions start when the feed contacts the hot regenerated catalyst in the riser and continues until the oil vapors are separated from the catalyst in the reactor. The hydrocarbon vapors are passed to multiprocessing (fractionating column, compression, absorption and distillation) to yield valuable product.

In this stage, the steam injected at 250 °C (523 K) in order to remove hydrocarbons adsorbed on internal and external surfaces of catalyst. Due to cracking reaction, coke formed on the catalyst surfaces that led to decrease the activation of catalyst therefore, the catalyst leaving the reactor is called spent catalyst contains coke deposited by the cracking, the spent catalyst enters the regenerator where coke is burned from the catalyst with air at the residence time is 10 seconds, supplied by the air compressor at 500 °C (773 K). The regenerator temperature and burned coke are controlled by varying the air flow rate. The catalyst then passes through the slide valve and enters into the bottom of the riser, thus forming a continuous catalyst circulation loop. Typical temperatures in the reactor is between 480-540 °C (753-813 K) where the reaction is endothermic while in the regenerator is between 650-815 °C (753-813 K) where the reaction is exothermic.

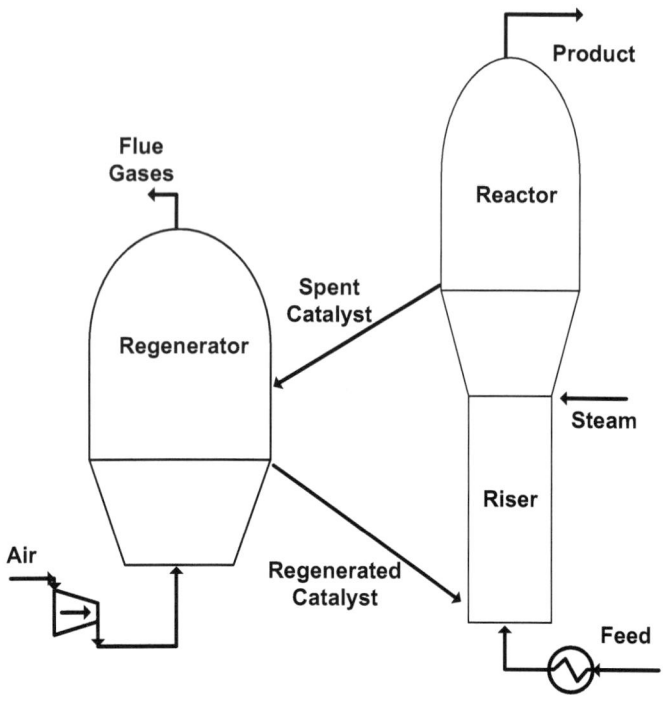

Figure (1.1): Flow sheet diagram of the side by side FCC Process.

Chapter Two: Dynamic Model based on Heat Balance

The mathematical model of various process units is derived using the equations of mass balance and/or energy balance as following:

Input – Output + Generation-Consumption = Rate of Accumulation.

In practice, application of this procedure introduces additional complexity into the system equation, so it is sometimes necessary to make simple assumptions of the dynamic behavior.

Mathematical models of chemical systems are developed for many reasons. Thus, they may be constructed to assist in the interpretation of experimental data, to predict the consequence of changes of system input or operating conditions, to deduce optimal system or operating conditions and for control purposes.

Usually there is an interest for dynamic model made to design and/or test the proposed control system. The dynamic and steady state simulation model of FCCU consist of a system of equations based on energy balances around the reactor and the regenerator.

1.1 Modeling of FCC Unit

Many researches in FCC unit relating to simulation, design, kinetics, dynamics and control have been published; these researches tried to give closed description in order to understand the nature of FCC unit process

The model used by Lee and Groves (1985) gives description in the fluid dynamic description of the regenerator. It is simple stirred tank with no dilute phase. The main problem of the model is a lack of detailed kinetics for the combustion of CO and CO_2 which occurs both at the solid catalytic surface and in the homogeneous phase. On the reactor side, Lee and Groves use the three-lump model.

McFarlane et al. (1993) published their which gives a very detailed and realistic description of the fluid dynamic behavior of the regenerator (bed density, distribution, flow, and catalyst density in the dilute phase) as well as a detailed description of the catalyst circulation in that specific reactor. It provides a comprehensive description of both the reactor and regenerator. Although the riser model is very simple, it provides sufficient information for regulatory control studies. Furthermore, the catalyst flow interactions between the reactor and regenerator are accounted for. Catalyst flow interactions between the reactor and regenerator sections are not addressed in any of the other models. All the models except the one by McFarlane et al. assume that the FCC unit is equipped with slide valves to control catalyst circulation rate.

Shaymaa H. Khazaal (2005) developed material and energy balance calculations to design FCC unit from Iraqi crude oil. Visual basic program was used in her work.

An integrated dynamic model for the complete description of the FCC unit was developed by Bollas G.M et al. (2007); the model simulates successfully the riser and the regenerator of FCC and incorporates operating conditions, feed properties and catalyst effects. The simulator can be utilized as a basis for a model based control of FCC units.

1.2 Control Formulation of FCC Unit

There are many control formulations that concern the description of the control system of fluidized catalytic cracking unit and how to choose the controlled variables and manipulated variables. The following are some scientific searches relating in this field.

Grosdidier (1988) et al. provided simulation results for a representative FCC unit regenerator control problem. The problem involves controlling flue gas composition, flue gas temperature, and regenerator bed temperature by manipulating feed oil flow, recycle oil flow and air to the regenerator.

McFarlane et al. (1993) developed their model where controlled variables are cracking temperature and flue gas oxygen concentration while manipulated variables are Lift air compressor speed and Flue gas valve opening. The control objective is to maintain the controlled variables (cracking temperature and flue gas oxygen concentration) at pre-determined set points in the presence of typical process disturbances while maintaining safe plant operation.

Moro & Odloak (1995) proposed six variables to be controlled: riser temperature, severity, temperature of the dense phase of the regenerator first stage, temperature of the regenerator dense phase of the second stage, the differential valve pressure of the regenerated catalytic and the rotation velocity of the gas compressor. They also proposed the following manipulated variables: feed flow rate to the unit, air flow rate to the regenerator, valve opening of the regenerated catalytic and the feed temperature. The chosen controlled variable here was the temperature of the dense phase of the regenerator first stage, manipulating the air flow rate to the regenerator.

Shinnar et al. (1996) selected regenerated flow rate and air flow rate as manipulated variables while reactor temperature and regenerator temperature as controlled variables where they show that this selection is more effective in control of FCC unit.

Uygen et al. (2006) applied Fuzzy logic control and model predictive control on FCC unit.

But in typical control of FCC unit, the manipulated variables are regenerated catalyst flow rate, reactor pressure and differential pressure between reactor and regenerator while controlled variable are reactor temperature, gas compressor and regenerator pressure.

In all these researches, many control strategies were proposed such as feedback control, advanced process control and even nonlinear control system.

Finally, the fluid catalytic cracking unit represents special problem in control system and it is difficult to reach to the optimal control system because of its dynamic is so complicated and varying according to typical design and operational requirements and tight of interaction between the reactor and regenerator.

1.3 Heat Balance of a FCC Unit

In FCC unit especially in the regenerator, the heat generated from combustion of coke is supplying total heat of all streams in the reactor and regenerator. In side of the reactor, it supplies the heat to raise the feed to rector temperature (200 to 500 °C), to substitute the heat losses by conduction, radiation, etc., to supply the heat of reaction (at 500 °C) and to raise the steam to reactor temperature (250 to 500 °C). While in the side of regenerator, it supplies the heat required to raise coke from reactor temperature to regenerator temperature (500 to 720 °C), To raise air to regenerator temperature (500 to 720 °C) and to substitute the heat losses by conduction, radiation, etc. the distribution of combustion heat is varying from stream to stream according to effect the stream on the process and design and operational consideration in order to maintain the good balances between reactor and regenerator. For example, the feed heating has the majority percentage of heat combustion while the heating air and the heat of reaction have moderate percentage and the heat losses and stream have the low percentage.

A number of simplified assumptions were made in order to formulate the energy balances in reactor and regenerator which include Neglecting the conduction, convection and radiation terms as well as Heat of reaction and Heat of combustion are constant.

1.3.1 Heat Balance around the Reactor

Heat of Input stream −Heat of Output stream + Heat of reaction=Rate of Accumulation

Heat of Regenerated catalyst + Heat of Feed + Heat of Steam − Heat of Product − Heat of Spent Catalyst + Heat of reaction= Rate of Accumulation

$$F_{rg}Cp_{rg}(T_{rg} - T_o) + F_fCp_f(T_f - T_o) + F_{st}H_{st} - F_pCp_p(T_{rc} - T_o) -$$
$$F_sCp_s(T_{rc} - T_o) + \Delta H_R = (M_s\,Cp_s + M_p\,Cp_p)\frac{dT_{rc}}{dt} \quad \text{....... (2.1)}$$

$\left(\text{oil feed + regCat + Steam} \right) - \left(\text{product r + Spent} \right) + M_R =$

1.3.2 Heat Balance around the Regenerator

Heat of Input stream − Heat of Output stream + Heat of reaction = Rate of Accumulation

Heat of Spent catalyst + Heat of Air − Heat of Combustion − Heat of Regenerated catalyst − Heat of Flue gases = Rate of Accumulation

$$F_s Cp_s(T_{rc} - T_o) + F_a Cp_a(T_a - T_o) - \Delta H_c - F_{rg} Cp_{rg}(T_{rg} - T_o) - F_f Cp_f(T_{rg} - T_o) =$$
$$\left(M_{rg}\, Cp_{rg} + M_f\, Cp_f\right) \frac{dT_{rg}}{dt} \quad\text{......}\ (2.2)$$

(spent + Air) − (regent + flue gas) Hc combs

Since, $M_s - M_{rg} = M_{coke}$ ...... (2.3)

In the side of the reactor, The Equation (2.1) will be:

$$F_{rg} Cp_{rg}(T_{rg} - T_o) + F_f Cp_f(T_f - T_o) + F_{st} H_{st} - F_p Cp_p(T_{rc} - T_o) -$$
$$F_{coke} Cp_{coke}(T_{rc} - T_o) + \Delta H_R = \left(M_{rg} Cp_{rg} + M_{coke} Cp_{coke} + M_p\, Cp_p\right) \frac{dT_{rc}}{dt} \quad\text{......}\ (2.4)$$

Or

$$F_{rg} Cp_{rg}(T_{rg} - T_{rc}) + F_f Cp_f(T_f - T_o) + F_{st} H_{st} - F_p Cp_p(T_{rc} - T_o) -$$
$$F_{coke} Cp_{coke}(T_{rc} - T_o) + \Delta H_R = \left(M_{rg} Cp_{rg} + M_{coke} Cp_{coke} + M_p\, Cp_p\right) \frac{dT_{rc}}{dt} \quad\text{......}\ (2.5)$$

In the side of the regenerator the Equation (2.2) will be:

$$F_{rg} Cp_{rg}(T_{rc} - T_o) + F_{coke} Cp_{coke}(T_{rc} - T_o) + F_a Cp_a(T_a - T_o) - \Delta H_c -$$
$$F_{rg} Cp_{rg}(T_{rg} - T_o) - F_f Cp_f(T_{rg} - T_o) = \left(M_{rg}\, Cp_{rg} + M_f\, Cp_f\right) \frac{dT_{rg}}{dt} \quad\text{......}\ (2.6)$$

Or

8

$$F_{coke}Cp_{coke}(T_{rc} - T_o) + F_aCp_a(T_a - T_o) - \Delta H_c - F_{rg}Cp_{rg}(T_{rg} - T_{rc}) -$$
$$F_fCp_f(T_{rg} - T_o) = (M_{rg}\,Cp_{rg} + M_f\,Cp_f)\,\frac{dT_{rg}}{dt} \quad \ldots\ldots (2.7)$$

Where:

Cp_{rg}: Specific heat of regenerated catalyst.

Cp_f: Specific heat of feed.

Cp_p: Specific heat of product.

Cp_s: Specific heat of spend catalyst.

Cp_a: Specific heat of air.

Cp_{fg}: Specific heat of flue gases.

F_{rg}: Mass flow rate of regenerated catalyst.

F_f: Mass flow rate of feed.

F_p: Mass flow rate of product.

F_s: Mass flow rate of steam. Spent entalyst

F_a: Mass flow rate of air.

F_{fg}: Mass floe rate of flue gases.

M_{rg}: Mass of regenerated catalyst.

M_p: Mass of reactor product.

M_{fg}: Mass of flue gases.

M_s: Mass of spend catalyst.

M_{coke}: Mass of coke.

T_{rg}: Temperature of regenerator.

T_f: Temperature of feed.

T_p: Temperature of product.

T_s: Temperature of steam.

T_a: Temperature of air.

T_{fg}: temperature of flue gases.

1.4 State Space Analysis

This method is widely used in modeling of dynamic system especially in multi-input and multi-output (MIMO) in order to carry out the transfer functions between input (manipulated and disturbance) variables and output (controlled) variables.

Calculation of this method, involves matrix algebra, depends on the group of matrices that defined in following equations:

$$\dot{x} = Ax + \begin{bmatrix} B_p & B_d \end{bmatrix} \begin{bmatrix} u \\ d \end{bmatrix} \quad ...... (2\ 8)$$

$$y = Cx + \begin{bmatrix} D_p & D_d \end{bmatrix} \begin{bmatrix} u \\ d \end{bmatrix} \quad ...... (2.9)$$

Where:

u = vector of manipulated variables.

y = vector of output variables.

d = vector of disturbances.

x = vector of system states.

A: is a matrix represents the coefficients of output variables (T_{rc} & T_{rg}).

B: is a matrix represents the coefficients of input variables (F_a & F_{rg}).

C: is an identity matrix.

D: is a matrix represents the coefficients of indirect variables. It is zero matrix expect in special process.

The elements of A matrix will be:

$$A_{ij} = \frac{\partial e_i}{\partial x_i} \quad ...... (2.10)$$

While the elements of B matrix will be:

$$B_{ij} = \frac{\partial e_i}{\partial u_i} \quad ...... (2.11)$$

Where: e_i refers to differential equation.

The state space model was used to convert the system from non-linear to linear in order to find the transfer functions between the manipulated/disturbance variables and controlled variables.

As in Equations (9) and (10) the state variables will be:

$$X = \begin{bmatrix} T_{rc} - T_{rcs} \\ T_{rg} - T_{rgs} \end{bmatrix} \dots\dots (2.12)$$

And the input variables are following:

$$U = \begin{bmatrix} F_{rg} - F_{rgs} \\ F_a - F_{as} \end{bmatrix} \dots\dots (2.13)$$

And the output variables are following:

$$Y = \begin{bmatrix} T_{rc} - T_{rcs} \\ T_{rg} - T_{rgs} \end{bmatrix} \dots\dots (2.14)$$

1.4.1 Calculation of matrix (A)

As in Equation (2.10), the matrix (A), with respect to this work, is defined as following:

$$A = \begin{bmatrix} A_{11} & A_{12} \\ A_{21} & A_{22} \end{bmatrix} \dots\dots (2.15)$$

Where:

11

$$A_{11} = \frac{\partial f_1}{\partial T_{rc}} = \frac{-F_p Cp_p - F_d Cp_d}{M_p Cp_p + M_d Cp_d} \ \ldots\ldots\ (2.16)$$

$$A_{12} = \frac{\partial f_1}{\partial T_{rg}} = \frac{F_d Cp_d}{M_p Cp_p + M_d Cp_d} \ \ldots\ldots\ (2.17)$$

$$A_{21} = \frac{\partial f_2}{\partial T_{rc}} = \frac{F_d Cp_d}{M_s Cp_s + M_f Cp_f} \ \ldots\ldots\ (2.18)$$

$$A_{22} = \frac{\partial f_2}{\partial T_{rg}} = \frac{-F_s Cp_s - F_f Cp_f}{M_s Cp_s + M_f Cp_f} \ \ldots\ldots\ (2.19)$$

1.4.2 Calculation of matrix (B)

As in Equation (2.11), the matrix (B), with respect to this work, is defined as following:

$$B = \begin{bmatrix} B_{11} & B_{12} \\ B_{21} & B_{22} \end{bmatrix} \ \ldots\ldots\ (2.20)$$

Where:

$$B_{11} = \frac{\partial f_1}{\partial F_{rg}} = \frac{Cp_{rg} T_{rg}}{M_p Cp_p + M_d Cp_d} \ \ldots\ldots\ (2.21)$$

$$B_{12} = \frac{\partial f_1}{\partial F_a} = \frac{0}{M_p Cp_p + M_d Cp_d} \ \ldots\ldots\ (2.22)$$

$$B_{21} = \frac{\partial f_2}{\partial F_{rg}} = \frac{-Cp_{rg} T_{rg}}{M_{rg} Cp_{rg} + M_f Cp_f} \ \ldots\ldots\ (2.23)$$

$$B_{22} = \frac{\partial f_2}{\partial F_a} = \frac{Cp_a T_a}{M_{rg} Cp_{rg} + M_f Cp_f} \ \ldots\ldots\ (2.24)$$

Chapter Three: Dynamics and Control

Multivariable control of FCC unit is complicated problem in selecting the appropriate manipulated and controlled variables. In this work, the Controlled variables were chosen to be the reactor temperature (T_{rc}), the regenerator temperature (T_{rg}) while the manipulated variables were chosen to be the regenerated catalyst flow rate (F_{rg}) and air flow rate (F_a). A schematic control diagram is shown in Figure (3.1).

The main aim of this work is to develop a mathematical model to control of the FCC unit, and simultaneous control of reactor temperature and regenerator temperature, subjected to regenerated catalyst flow rate and air flow rate respectively.

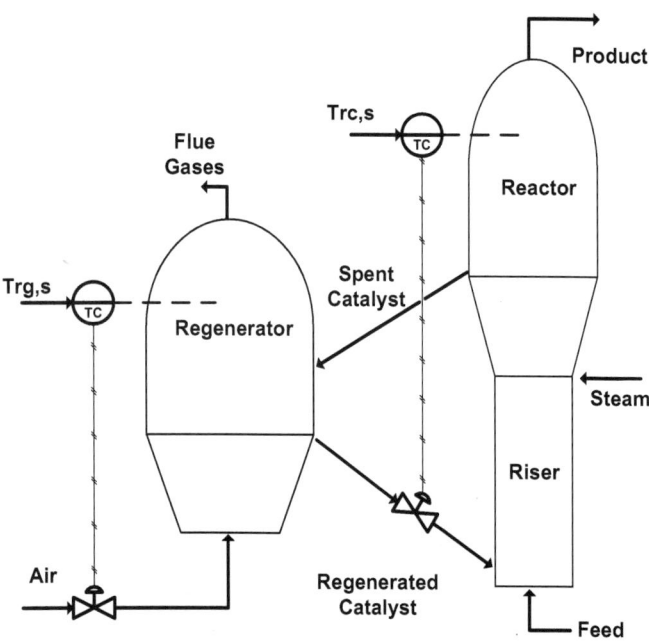

Figure (3.1): Schematic control diagram of FCCU.

3.1 Strategies of FCC unit Control

One of the important considerations in design of control system is the nominee the control schemes therefore; there are four schemes (A, B, C, D) used in this work.

✓ Scheme A (MIMO system) refers to PID Feedback control system where interaction between the manipulating variables and controlled variables was considered.

✓ Scheme B refers to PID Feedback with two decouplers control systems to cancel the interaction between the manipulating variables (F_{rg} & F_a) and the controlled variables (T_{rc} & T_{rg}) in reactor and regenerator.

✓ Scheme C (MIMO system) refers to Fuzzy logic control system where interaction between the manipulating variables and controlled variables was considered.

✓ Scheme D refers to Fuzzy logic control with two decouplers control systems to cancel interaction between the manipulating variables (F_{rg} & F_a) and controlled variables (T_{rc} & T_{rg}) in the reactor and regenerator.

3.2 Decoupling Control System

The interaction in the process happened when one or more manipulated variable(s) has/have affect on two or more controlled variables, in this case, the relative array should be calculated to find the tight of interaction and applying the decoupling system.

3.2.1 Relative Gain Array (RGA)

RGA is a matrix of gains where the value of gains in rows or columns is equal to one. This matrix is almost square but sometime is non-square matrix depending on the number of selected manipulated and controlled variables in the process. Now, it is appropriate method which gives acknowledgment in tight of interaction in order to choosing the best pair of manipulated and controlled variables.

Calculation of RGA depends on the gains of the open loop and closed loop at steady state, when PI(D) feedback control is applied, it depends only on the gains of open loop.

The aim of the calculation is to couple the output (controlled) variables with input (manipulated) variables that have the largest gains in matrix.

In this work, it is convenient to arrange a table of those gains in form of a matrix, the relative gain array of interacting process is:

$$\Lambda = \begin{matrix} F_{rg} & F_a \\ \begin{bmatrix} \lambda_{11} & \lambda_{12} \\ \lambda_{21} & \lambda_{22} \end{bmatrix} & \begin{matrix} T_{rc} \\ T_{rg} \end{matrix} \end{matrix} \quad \ldots\ldots (3.1)$$

Relationships of two controlled outputs and two manipulated inputs as shown are given by:

$$T_{rc}(s) = H_{11}F_{rg}(s) + H_{12}F_a(s) \ldots\ldots (3.2)$$

$$T_{rg}(s) = H_{21}F_{rg}(s) + H_{22}F_a(s) \ldots\ldots (3.3)$$

Where $H_{11}(s)$, $H_{12}(s)$, $H_{21}(s)$ and $H_{22}(s)$ are the four transfer functions relating the two outputs (T_{rc} and T_{rg}) to the two inputs (F_{rg} and F_a). Equations (24) and (25) indicate that a change in (F_{rg} or F_a) will affect both controlled outputs.

The relative gain between the controlled variable (T_{rc}) and the manipulated variable (F_{rg}) will be denoted by (λ_{11}).

$$\lambda_{11} = \frac{\text{Open loop gain}}{\text{closed loop gain}} \quad \ldots\ldots (3.4)$$

Mathematically, the relative gain can be expressed as:

$$\lambda_{11} = \frac{(\Delta T_{rc} / \Delta F_{rg})_{F_a}}{(\Delta T_{rc} / \Delta F_{rg})_{T_{rg}}} = \frac{H_{11}}{\frac{H_{11}H_{22}-H_{12}H_{21}}{H_{22}}} \quad \ldots\ldots (3.5)$$

15

$$\lambda_{11} = \frac{1}{1 - \frac{H_{12}H_{21}}{H_{11}H_{22}}} \quad \dots \quad (3.6)$$

.

In most cases, the steady state relative gain analysis is a sufficient indicator for control loops combination, so the steady state relative gain could be calculated from the following equations:

$$\lambda_{11} = \frac{1}{1 - \frac{K_{12}K_{21}}{K_{11}K_{22}}} \quad \dots \quad (3.7)$$

The open loop static gain is between (T_{rc} and F_a) when (F_{rg}) is kept constant and the other, when (T_{rg}) is constant by the control loop. The values of other relative gains could be calculated from above relative gain (λ_{11}).

Where: $\lambda_{12} = \lambda_{21} = 1 - \lambda_{11}$, $\lambda_{11} + \lambda_{12} = 1$ and $\lambda_{11} + \lambda_{21} = 1$

The RGA provides a useful measure of interaction. In particular:

✓ If $\lambda_{11} = 0$, then T_{rc} does not respond to F_{rg} and F_{rg} should not be used to control T_{rc}.

✓ If $\lambda_{11} = 1$, then F_a does not affect T_{rc} and the control loop between T_{rc} and F_a does not interact with loop of T_{rg} and F_a. In this case we have completely decoupled loops.

✓ If $0 < \lambda_{11} < 1$, then an interaction exists and as F_a varies it affects the steady-state value of T_{rc}. The smaller the value of λ_{11}.the larger the interaction becomes.

✓ If $\lambda_{11} < 0$, then F_a causes a strong effect on T_{rc} and in opposite direction from that caused by F_{rg}. In this case, the interaction is very dangerous.

In other hand, the elements of the RGA can be calculated for a system of any size (more than 2×2 size and matrix should be square) using the following equation.

$\lambda_{ij} = $ (ijth element of k_p) × (ijth element of $[k_p^{-1}]^T$) $\quad \dots \quad (3.8)$

Where: K_p is the gain of the open loop system.

3.2.2 Design of Non-interacting Control Loops

The benefit of using decouples is cancelling the interaction effect between the two loops and thus rearrange it to two non-interacting control loops.

To design decouples for a FCC unit, Equations (24) and (25) have been used. From Equation (3.23), in order to keep T_{rc} constant (i.e. $T_{rc}=0$), F_s should be changed by the following:

$$0 = H_{11}F_{rg}(s) + H_{12}F_a(s) \ldots\ldots (3.9)$$

$$F_{rg} = -\frac{H_{12}(s)}{H_{11}(s)} F_a(s) \ldots\ldots (3.10)$$

Equation (3.10) implies that dynamic element is introduced with a transfer function:

$$D_1(s) = -\frac{H_{12}(s)}{H_{11}(s)} \ldots\ldots (3.11)$$

It uses the value of F_a as input and provides as output the amount by which it should change F_{rg}, in order to cancel the effect of F_a on T_{rc}.

This dynamic element (decoupler) when installed in the control system cancels any effect that loop 2 might have on loop 1, but not vice versa.

To eliminate the interaction from loop 1 and loop 2, the same reasoning as above has been followed and it was found that the transfer function of the second decoupler is given by:

$$D_2(s) = -\frac{H_{21}(s)}{H_{22}(s)} \ldots\ldots (3.12)$$

From previous equations, the following two closed loop input-output relationships are developed as:

17

$$T_{rc} = \frac{G_{c1}[H_{11}-H_{12}H_{21}/H_{22}]}{1+G_{c1}[H_{11}-H_{12}H_{21}/H_{22}]} T_{rc,sp} \quad \ldots \ldots \text{(3.13)}$$

$$T_{rg} = \frac{G_{c2}[H_{22}-H_{12}H_{21}/H_{11}]}{1+G_{c2}[H_{22}-H_{12}H_{21}/H_{11}]} T_{rg,sp} \quad \ldots \ldots \text{(3.14)}$$

Where $T_{rc,sp}$ and $T_{rg,sp}$ are the set point value of T_{rc} and T_{rg}, respectively G_{c1} and G_{c2} are the controller transfer functions of the first and second loops respectively. The last two equations demonstrate complete decoupling of the two loops.

3.3 PID Feedback Controller

In temperature control, PID Feedback is always applied. The reasons are gathering between advantages of integral action where offset is equal to zero and advantages of derivative action where it reduces the oscillation.

It can be expressed mathematically as:

$$G_c = K_c + \frac{K_c}{\tau_I s} + K_c \tau_D s \quad \ldots \ldots \text{(3.15)}$$

The specification of closed loop response depends on the suitable selection tuning of the controller for appropriate values of K_C, τ_I and τ_D to give a stable system.

3.3.1 Control Tuning

In Feedback control, the essential topic is selecting best method in control tuning that is mean best control parameters (K_C, τ_I and τ_D). There are several methods; some of it depends on open loop behavior and other depends on closed loop behavior. In this work, two methods were chosen to find the initial values of K_C, τ_I and τ_D.

These methods are:

1. Process Reaction Curve (PRC).

2. Internal Model Control (IMC)

With respect to process reaction curve (PRC), this method, proposed by Cohn and Coon in 1953, depends on specifications of response when a step change in the manipulated variable is happened. The object is to convert the dynamic response to the first order transfer function with lag time without requiring detailed knowledge of the process as following:

$$G_{prc} = \frac{Ke^{-t_d s}}{\tau s + 1} \quad \text{....... (3.16)}$$

The values of K, τ and t_d are calculated from the process reaction curve which is shown in Figure (3.2) as following:

$$K = \frac{B}{A} = \frac{Output\ at\ steady\ state}{Input\ at\ staedy\ state} \quad \text{....... (3.17)}$$

$$\tau = \frac{B}{S} = \frac{Output\ at\ steady\ state}{Slope\ of\ response\ curve} \quad \text{....... (3.18)}$$

t_d is required time until the system responds

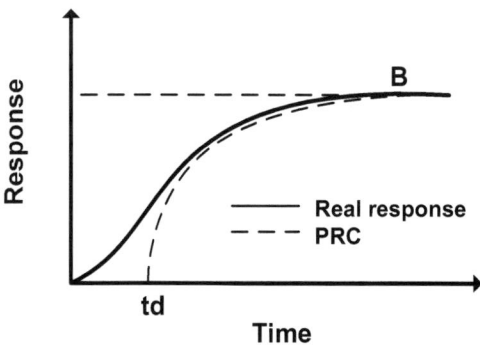

Figure (3.2): Approximation of process reaction curve method.

Since the PID Feedback was chosen, the control parameter is calculated as following:

$$K_c = \frac{1}{K} \frac{\tau}{t_d} \left(\frac{4}{3} + \frac{t_d}{4\tau}\right) \dots\dots (3.19)$$

$$\tau_I = t_d \frac{32 + 6 \, ^{t_d}/_\tau}{13 + 8 \, ^{t_d}/_\tau} \dots\dots (3.20)$$

$$\tau_D = t_d \frac{4}{11 + 2 \, ^{t_d}/_\tau} \dots\dots (3.21)$$

With respect to the internal control method (IMC), initial values for PID feedback systems are shown in following:

$$K_c = \frac{1}{K_p} \frac{2\tau_p/t_d + 1}{2\tau_c/t_d + 1} \dots\dots (3.22)$$

$$\tau_I = \tau_p + t_d/2 \dots\dots (3.23)$$

$$\tau_I = \frac{\tau_p}{2\tau_p/t_d + 1} \dots\dots (3.24)$$

ISE criteria was used to carry out the best initial values of the controller parameters (K_C, τ_I and τ_D) using PRC and IMC methods and to compare between interaction and decoupling systems. ISE means area under the curve in plotting between the squared error and time. It is defined as:

$$ISE = \int_0^t E^2 (t)dt \dots\dots (3.25)$$

Where: E is error

In Feedback control, Error (t) = Set point (t) – Measured values (t).

3.4 Fuzzy Logic Control (FLC)

Fuzzy control system considers one of advanced control systems, was invented by Lotfi A. Zadeh in 1965, it is super set of traditional logic. The theory of Fuzzy sets has one of its aims, the development of a methodology for the formulation and solution of problems that are too complex or too ill-defined to be analyzed by conventional techniques. Hence the theory of Fuzzy sets is likely to be recognized as a natural development in the evaluation of scientific thinking. The benefits of fuzzy logic are more than conventional control, where it is cheaper to make and easier to design, it simplifies knowledge acquisition and representation, it, therefore it gives more accurate results.

3.4.1 Design of Fuzzy Logic Controller

The aim of any process controller is relating the state variables to action variables. Now, A Fuzzy logic controller constructed to implement the known heuristic. Thus in such a controller the variables are equated to non-Fuzzy universe given the possible range of measurement or action magnitudes. These variables, however, take on linguistic values which are expressed as Fuzzy subset of the universe. The design of Fuzzy logic controller can be describe as following:

✓ Choosing a suitable scaled universe of discourse $(_\upsilon)$ in range of $-L \leq (E_i, CE_i) \leq L$,

Where: L and $-L$ represent to the positive and negative ends respectively of this universe. E_i and CE_i represent the error and its rate of change for the same instant (i).

✓ Defining the non-Fuzzy set intervals (the quantized levels scaled values) for E_i, CE_i and control action (U). Each level has a value (I) lying between $(-XG \leq I \leq XG)$ where: XG and $-XG$ represent the controller gain and they are regarded as the values of the universe of discourse limits (L and $-L$) respectively.

✓ Defining Fuzzy-sets definitions in control for E, CE and U to have these forms:

PB=positive Big   NB=Negative Big   Z=Zero

PS=Positive Small   NS=Negative Small

✓ Choosing the appropriate rule Fuzzy logic controller (9, 25 or 49 rules). The Fuzzy decision rules are developed linguistically to do a particular control action and are arranged in form of set of Fuzzy conditional statements as following:

IF E IS NB AND CE IS NB THEN PB ACTION

This form can be rearranged in other form with the help of Fuzzy sets definition into a new statement.

IF NEB AND NCB THEN PUB

Table (3.1): 25-Rule Fuzzy Logic Controller.

	NCB	NCS	ZC	PCS	PCB
NEB	PUB	PUB	PUB	PUS	ZU
NES	PUB	PUS	PUS	ZU	NUS
ZE	PUB	PUS	ZU	NUS	NUB
PES	PUS	ZU	NUS	NUS	NUB
PEB	ZU	NUS	NUB	NUB	NUB

Table (3.1) shows the Fuzzy rules conclusions. The five Fuzzy sets definition generates (25) rules Fuzzy controller. To read these one can obtain the following translation of the first three rules.

IF NEB AND NCB THEN PUB
IF NEB AND NCS THEN PUB
IF NEB AND ZC THEN PUB...and so on

It is important to know that in any control system always the error and change of error define as:

E_i = (Set values)$_i$ – (Measured values)$_i$ ...... (3.26)

CE_i = (Instant error)$_i$ – (pervious error)$_i$ ...... (3.27)

But, in Fuzzy logic control define as:

E_i = (Measured values)$_i$ - (Set values)$_i$ ...... (3.28)

To clear this difference, consider the initial condition state for a system subjected to a unit step change in input.

For Feedback controller,

$E(0) = +1$, $CE(0) = 0$

And according to Table (3.1), the Fuzzy rule will be:

IF BEB AND ZC THEN NUB

The action will be negative and the output will follow it. To solve this problem, Equation (3.28) should be used.

$E(0) = -1$, $CE(0) = 0$

And the Fuzzy rule will be:

IF NEB AND ZC THEN PUB

So the action will be positive and the output will follow it. This Fuzzy definition E and CE will be considered in this work.

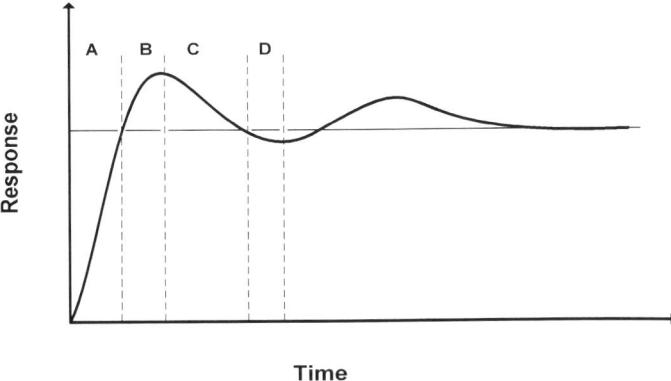

Figure (3.3): Transient response of closed loop.

According to Equations (3.27 & 3.28) and Figure (3.3) the signs of error and the change of error will be as following in Table (3.2):

Table (3.2): Signs Distribution in Fuzzy logic controller

Section	Signs of error	Signs of change of error
A	-	+
B	+	+
C	+	-
D	-	-

✓ Multiplying Both E_i and CE_i by the scale factor of the universe of discourse to ensure mapping their values into suitable intervals that belong to each one, also this scale factor helps to simplify handling of the numerical values of all variables.

✓ Calculation scalar control action (U_s), using the center of gravity method as flowing:

$$U_s = \frac{\sum_{n=1}^{N} \mu_n u_n}{\sum_{n=1}^{N} \mu_n} \quad \text{......} \ (3.29)$$

Where: (μ) represents the elements (membership) of the net control action vector. (u) represents the value on the interval n. An integral procedure (an algebraic sum) is required to obtain the effective control action scalar for each instant (i).

$$U_{s_{i+1}} = U_{s_i} + U_{s_{i-1}} \quad \text{......} \ (3.30)$$

A scalar factor is used to remove the first scalar factor in order to put the values into real one.

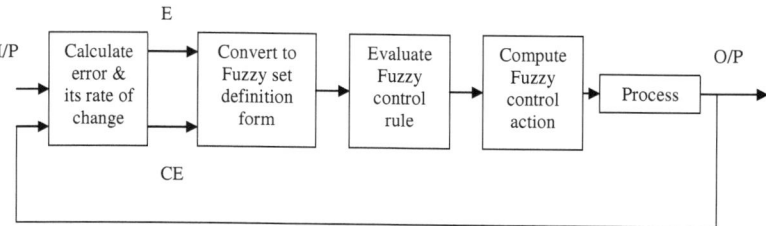

Figure (3.4): Block diagram of a control system using Fuzzy logic control.

3.4.2 Fuzzy Logic Control Procedure for MIMO System

Figure (3.4) describes a 2×2 Fuzzy controlled process. The Fuzzy control procedure for MIMO system is similar to the one for SISO process. All Fuzzy control functions are defined and calculations are made except that the Fuzzy rules will be divided for each controlled variable taking into account the other controlled variables with (ANY membership) which gives a membership ($\mu = 1$) whenever it appears. To clarify the idea, the following Fuzzy rules are examined:

IF E1 IS PEB AND CE1 IS PCS AND E2 IS ANY AND CE2 IS ANY THEN NUB

The same shape of rules will be fulfilled for other controlled variable as shown below:

IF E1 IS ANY AND CE1 IS ANY AND E2 IS PEB AND CE2 IS PCS THEN NUB

And so on for all rules. From the definition of AND (min), (ANY) membership will have no effect on the control procedure.

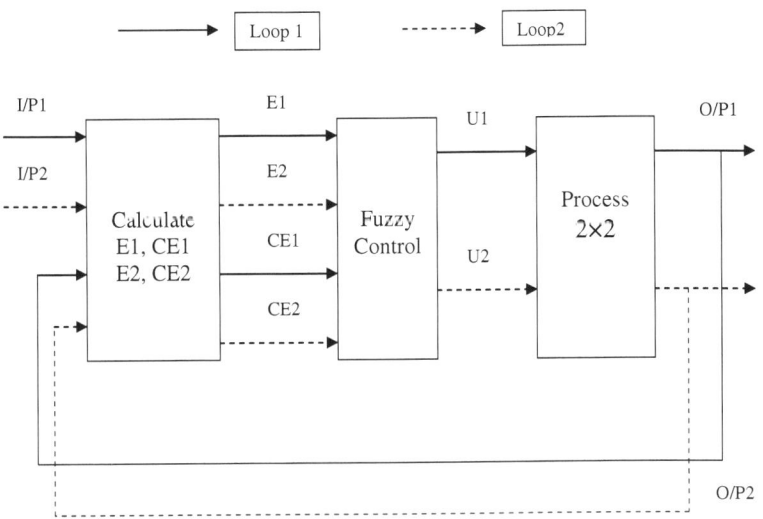

Figure (3.5): block diagram of Fuzzy logic controller for 2×2 process.

3.4.3 Fuzzy Control Tuning and Selection

There are three parameters should be accounted to tune of fuzzy logic. The parameters are:

✓ Gain tuning: depends on varying the gain and fixing other parameters.
✓ Interval tuning: depends on varying the interval and fixing other parameters.
✓ Fine tuning: depends on using more than one digit.

In addition, to choose the suitable controller, the controller ability to give a reasonable response, this depends on Number of rules, Number of interval, Interval values and Fuzzy sets definition

Chapter Four: Results and Discussion

A good starting point in modeling of FCC unit is the good selection of the manipulated and controlled variables in reactor and regenerator. In this work, dynamic models have been developed to study the influence of manipulated variables on controlled variables of the process.

This chapter illustrates the results obtained from the computer programs in chapter five for different control strategies and schemes. The results including two main parts:

The first part was to study the dynamic behavior of the system where the transfer functions between the controlled variables and manipulated variables were computed using state space model then plotting the step response.

The second part was to study the control behavior which is the main aim of this work by applying different control strategies and schemes of FCC unit.

In the control behavior, PID Feedback controller mode and Fuzzy logic control were applied; Integral of Square Error (ISE) was carried out for each scheme (A, B, C and D) for the comparison among them.

4.1 Analysis of the Dynamic behavior

4.4.1 Results of State Space model

The simulation of state space model is calculated from chapter five, and the results are:

$$A = \begin{bmatrix} A_{11} & A_{12} \\ A_{21} & A_{22} \end{bmatrix} = \begin{bmatrix} -0.2 & 0.119 \\ 0.079 & -0.1 \end{bmatrix} \ \dots\dots \ (4.1)$$

$$B = \begin{bmatrix} B_{11} & B_{12} \\ B_{21} & B_{22} \end{bmatrix} = \begin{bmatrix} 0.187 & 0 \\ -0.094 & 0.156 \end{bmatrix} \ \dots\dots \ (4.2)$$

$$C = \begin{bmatrix} 1 & 0 \\ 0 & 1 \end{bmatrix} \ \dots\dots \ (4.3)$$

27

$$D = \begin{bmatrix} 0 & 0 \\ 0 & 0 \end{bmatrix} \quad \text{....... (4.4)}$$

Where:

A: is a matrix represents the coefficients of state variables (T_{rc}, T_{rg}).

B: is a matrix represents the coefficients of input variables (F_a, F_{rg}).

C: is an identity matrix.

D: is a matrix represents the coefficients of indirect variables.

4.4.2 Dynamic Behavior

Figure (4.1) shows the dynamic responses were studied for step changes in the manipulated variables (F_{rg} and F_a) in order to study the effect of each change on the controlled variables (T_{rc} and T_{rg}) while Figure (4.2) refers to the deviation response. It is so clear the effects of the regenerated catalyst flow rate and air flow rate on the reactor temperature and regenerator temperature.

Upper left Figures (4.1 & 4.2) shows the effect of regenerated catalyst flow rate on the reactor temperature where reactor temperature increases and then rapidly reaches to steady state.

Upper right Figures (4.1 & 4.2) shows the effect of air flow rate on the reactor temperature where reactor temperature increases but required time to reached to steady state.

Lower left Figures (4.1 & 4.2) shows the effect of regenerated catalyst flow rate on the regenerator temperature, it is clear regenerator temperature decreases when positive step change in regenerated catalyst flow rate.

Lower right Figures (4.1 & 4.2) shows the effect of air flow rate on the regenerator temperature. It is proportional relation.

The response to reach steady state in the reactor temperature is faster than in the regenerator temperature to a step change in the regenerator catalyst flow rate But similar to a step change in the air flow rate. The transfer functions that describe the dynamic behavior from state space are following:

28

$$T_{rc}(s) = \frac{17.7s+0.7}{94.5s^2+28.35+1}F_{rg}(s) + \frac{1.766}{94.5s^2+28.35+1}F_a(s) \ \ldots\ldots\ (4.5)$$

$$T_{rg}(s) = \frac{-8.9s-0.38}{94.5s^2+28.35+1}F_{rg}(s) + \frac{14.8s+2.96}{94.5s^2+28.35+1}F_a(s) \ \ldots\ldots\ (4.6)$$

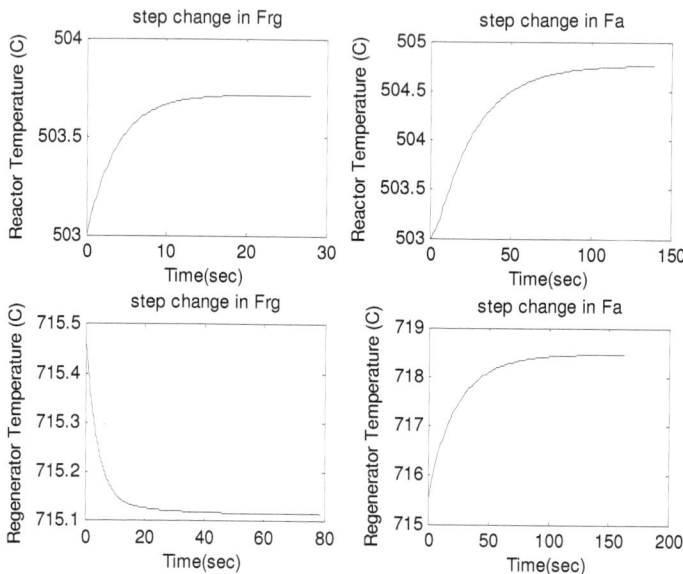

Figure (4.1): Dynamic behavior in MIMO system.

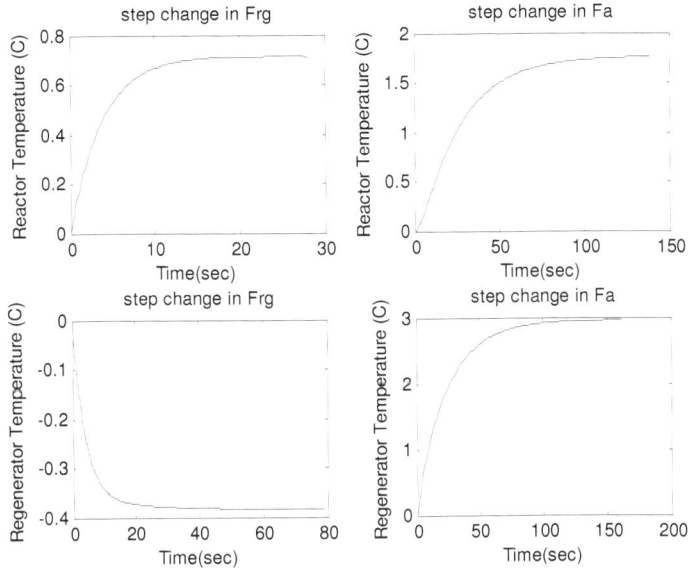

Figure (4.2): Deviation response in MIMO system.

To study the interaction between these variables, a step change in the (F_{rg} & F_a) should be applied in same time to carry out the effects on the (T_{rc} & T_{rg}) to find the approximation of system parameters (K, τ and t_d) for control tuning. The results are in Table (4.1) and Figure (4.3).

Table (4.1): System parameters for step changes in both F_{rg} & F_a.

Process	K	τ	t_d
Reactor	2.46	22	2
Regenerator	2.56	23	3

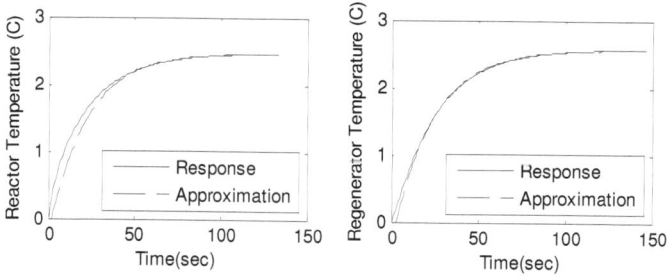

Figure (4.3): Approximation responses in both reactor and regenerator.

4.4.3 Relative Gain Array (RGA) Calculations

RGA must be calculated to choose the best pairing of the two controlled variables (T_{rc} and T_{rg}) and the two manipulated variables (F_{rg} and F_a) before applying the control techniques. The resulted array is given as:

$$RGA = \begin{bmatrix} \lambda_{11} & \lambda_{12} \\ \lambda_{21} & \lambda_{22} \end{bmatrix} = \begin{bmatrix} 0.75 & 0.25 \\ 0.25 & 0.75 \end{bmatrix} \quad \dots\dots (4.7)$$

So the best coupling are obtained by pairing the reactor temperature (T_{rc}) with regenerated catalyst flow rate (F_{rg}), and the regenerator temperature (T_{rg}) with air flow rate (F_a), since λ_{11} is positive and greater than 0.5.

4.5 Analysis of Control Behavior

In this section two different control strategies were used, the PID Feedback control and Fuzzy logic control, these control strategies were applied to four different schemes A, B, C and D.

In Feedback system, PID controller mode was used to control the temperatures of both reactor and regenerator; therefore, tuning the control parameters (proportional gain (K_c), time integral (τ_I) and time derivative (τ_D)) must be applied.

4.4.1 Results of control tuning

Comparison between two methods depends on the transient response of closed loop, when those methods were used, such as overshoot, decay ration, rise time, settling time and ISE. In this work, Settling time and ISE were selected in doing comparison where the process has high velocities in regenerated catalyst flow rate and air flow rate which means the reaching to steady state in an important aspect.

The initial values of the controller parameters $(K_c, \tau_I$ & $\tau_D)$ were tuned using computer simulation programs based on minimum integral of square error (ISE) and settling time are given in Table (4.2-4.3).

The control tuning was found in two different methods therefore; it can be seen that the tuning using the process reaction curve method is better than internal model control because the ISE value of first method is less than second method. In addition, PRC gives lower settling time than IMC.

Also it is clear that the scheme B is better than scheme C as shown their ISE values tabulated above therefore; it can be concluded that the two way decoupling system is better than interaction system.

Table (4.2): Control parameters of PID controller of reactor.

Control Tuning Methods	Control Parameters			ISE		Settling Time (sec)	
	K_c	τ_I	τ_D	Scheme A	Scheme B	Scheme A	Scheme B
PRC	6.03	4.74	0.71	0.2722	0.3929	20	5
IMC	3.98	23	0.95	0.962	0.5488	20	18

Table (4.3): Control parameters of PID controller of regenerator.

Control Tuning Methods	Control Parameters			ISE		Settling Time (sec)	
	K_c	τ_I	τ_D	Scheme A	Scheme B	Scheme A	Scheme B
PRC	4.07	7	1.06	0.45	0.77	25	14
IMC	2.72	24.5	1.40	1.15	2.63	25	16

4.2.2 Control Behavior

Figure (4.4) represents the closed loop behavior in reactor for both interaction system (scheme A) and decoupling system (Scheme B) using different methods (PRC & IMC).

It is so clear, the effect the interaction on system where the response has oscillation and required long time to reach to steady state, in other hand the decoupling system has soft response and faster time to steady state, the same controller parameters in both schemes A & B were applied on the reactor. Another comparison is between PRC and IMC, PRC is better than IMC for both cases.

Figure (4.5) represents the closed loop behavior in regenerator for both interaction (scheme A) and decoupling system (Scheme B) using different methods (PRC & IMC).

The results are the same relating to the interaction and decoupling systems, but it is so clear that the response in the reactor is faster than in the regenerator.

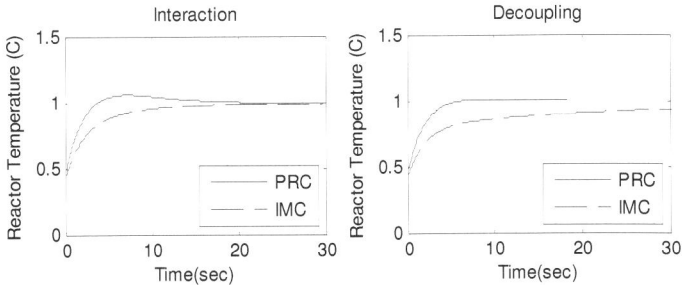

Figure (4.4): Control behavior of reactor using PID Feedback.

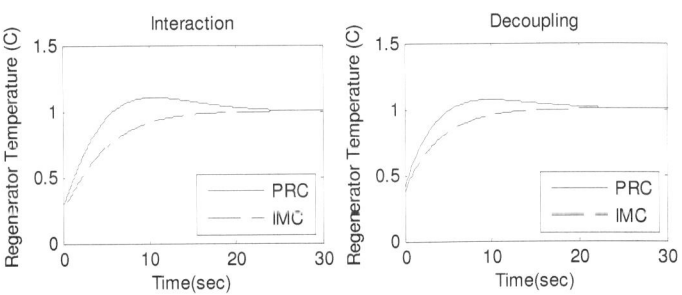

Figure (4.5): Control behavior of regenerator using PID Feedback.

4.4.3 Decoupler Design

The decoupler of loop1 ($D_1(s)$) was designed to eliminate the effect of interaction of loop2 on loop1 using Equation (3.11) on substitution the values of $H_{11}(s)$ and $H_{12}(s)$ the decoupler shows the following value:

$$D_1(s) = -\frac{H_{12}}{H_{11}} = -\frac{1.76}{17.7s+0.7} \cdots\cdots (4.8)$$

The value of $D_1(s)$ is coupled with the value of the main regenerated catalyst flow rate (F_{rg}) to get the final value, after each time interval.

In the same way, the decoupler of loop2 ($D_2(s)$) was designed to eliminate the effect the loop1 on loop2. After applying the value of $H_{21}(s)$ and H_{22} (s), the value is:

$$D_2(s) = -\frac{H_{21}}{H_{22}} = \frac{8.9s+0.38}{14.8s+2.96} \cdots\cdots (4.9)$$

The decoupler obtained to justify the main value of air flow rate (F_a).

4.3 Fuzzy Logic Controller (FLC) Behavior

4.3.1 Controller Gain Tuning

The control tuning of the FLC has two methods, first method depends on equations of classical control parameters (K_c, τ_I & τ_D) to find the value of the controller gain (XG) and second method depends on trial and error to find the value of the controller gain (XG) therefore, selection between two method depends on experience of researcher or engineer and nature of process. Tuning of FLC was used in two schemes C and D. The values of the controller gains were tuned using computer simulation program based on minimum integral of square error (ISE) and settling time.

The results of the controller gain of schemes C and D are given in Tables (4.4 & 4.5) of the reactor and regenerator respectively.

Figures (4.6) shows the control response using trial and error tuning method of reactor for interaction system (scheme C) and decoupling system (scheme D) respectively.

Figures (4.7) shows the control response using trial and error tuning method of regenerator for interaction system (scheme C) and decoupling system (scheme D) respectively.

In this work, comparison between two schemes depends on two factors ISE criteria and settling time, selecting or focusing on these two factors is popular in control process as well as overshoot, 1/4 decay ratio, and rise time to carry out the appropriate responses of the systems. As known, the process has high velocities flow rate, therefore high response in the controlled variables but, in same time ISE should be accounted to reach, as possible, the best value of tuning.

Table (4.4): Controller gains of reactor in FLC.

Controller Gain	ISE		Settling Time (sec)	
	Scheme C	Scheme D	Scheme C	Scheme D
50	0.06	0.06	4	2

Table (4.5): Controller gains of regenerator in FLC.

Controller Gain	ISE		Settling Time (sec)	
	Scheme C	Scheme D	Scheme C	Scheme D
60	0.05	0.10	2	2

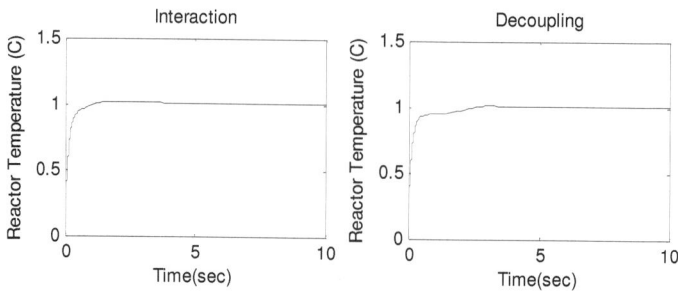

Figure (4.6): Control behavior of reactor using Fuzzy logic.

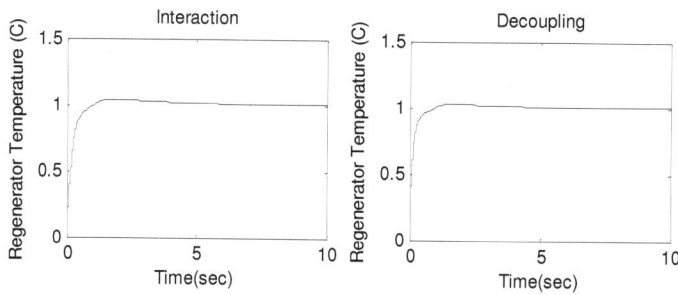

Figure (4.7): Control behavior of regenerator using Fuzzy logic.

PID controller was considered for comparison study with FLC because it is still the widely used strategy in industry. To make a clear comparison between these controllers, ISE and settling time were used as an index. In this comparison, all controllers (FLC and PID Controller) were tuned to the approximately best settings. In general FLC gives better results than PID controller, where the advantage of the FLC is that it does not need a model to build the control settings as in the case of PID controller.

Chapter Five: Simulation

5.1 MATLAB programs

In this chapter, the computer program were developed for the dynamic and control behavior using MATLAB version 7.10 (R2010a) program. Each program was executed and the results were checked to meet the model requirements.

MATLAB (Matrix Laboratory) program has proved high activity in many fields especially in modeling, simulation, control and numerical applications. Screen windows functions built-in this program and are fixable and easy to learn and solve the equations, matrix algebra and ... etc). To build program, it is important to know using M-file and appropriate functions because is the key for all mathematical calculation and displaying results and plots.

Table (5.1) lists some functions and commands and their description that were used in computer simulation.

Function Name	Function Description
axis	Specific the manual axis scaling on plot
feedback	Computes the feedback interconnection of tow systems
grid on	Add the grin to the current graph
hold on	Holds the current graph on the screen
legend	Puts a legend on the current screen
plot	Generates a linear plot
series	Computes a series system connection
ss	Creates a state-space model object
step	Calculates a unit step response of a system
tf	Creates a transfer function model object
title	Add a title to the current graph
trapz	Computes the integration value
xlabel	Add the label to the x-axis of the current graph
ylabel	Add the label to the y-axis of the current graph

Table (5.1) Summary functions in MATLAB program.

38

Steps of MATLAB program using M-file for dynamic behavior as following:

```
%MATLAB Program for dynamic behavior
clear all, clc
%Define Matrices A, B, C and D
%Matrix A
%A=[a11   a12; a21 a22];
%Matrix B
%B=[b11  b12; b21+0.025  b22];
%Matrix C
%C=[1  0; 0  1];
%Matrix D
%D=[0  0; 0  0];
A=[-0.2  0.119; 0.079 -0.1];
B=[0.187  0; -0.094  0.156];
C=[1  0; 0  1];
D=[0  0;0  0]'
%Transfer function between input Frg with both outputs (Trc & Trg)
[num1,den]=ss2tf(A,B,C,D,1)
%Transfer function between input Fa with both outputs (Trc & Trg)
[num2,den]=ss2tf(A,B,C,D,2)
%
%calculation the Relative gain array(RGA)
%Define steady-state gain matrix
numfrg=num1./den(3);
numfa=num2./den(3);
denn=den./den(3);
k11=numfrg(1,3);
k12=numfa(1,3);
k21=numfrg(2,3);
```

```
k22=numfa(2,3);
lamda11=1/(1-((k12*k21)/(k11*k22)));
RGA=[lamda11   1-lamda11;1-lamda11  lamda11]
%or using another method
kp=[k11 k12;k21 k22];
%calculate matrix inverse then take its transpose
%Multiplication element by element using ".*" operator
kpinverse=inv(kp);
kptranse=inv(kp)';
RGA=kp.*kptranse
%
%Trasfer Function
H11=tf(numfrg(1,:),denn)
H12=tf(numfa(1,:),denn)
H21=tf(numfrg(2,:),denn)
H22=tf(numfa(2,:),denn)
%
Trcss=503;%Reactor temperature at steady state
Trgss=715.5;%Regenerator temperature at steady state
figure(1)
[yh11,xh11,th11]=step(numfrg(1,:),denn);
[yh12,xh12,th12]=step(numfa(1,:),denn);
[yh21,xh21,th21]=step(numfrg(2,:),denn);
[yh22,xh22,th22]=step(numfa(2,:),denn);
subplot(2,2,1)
plot(th11,yh11+Trcss,'k-')
xlabel('Time(sec)')
ylabel('Reactor Temperature (C)')
title('step change in Frg')
%
```

```
subplot(2,2,2)
plot(th12,yh12+Trcss,'k-')
xlabel('Time(sec)')
ylabel('Reactor Temperature (C)')
title('step change in Fa')
%
subplot(2,2,3)
plot(th21,yh21+Trgss,'k-')
xlabel('Time(sec)')
ylabel('Regenerator Temperature (C)')
title('step change in Frg')
%
subplot(2,2,4)
plot(th22,yh22+Trgss,'k-')
xlabel('Time(sec)')
ylabel('Regenerator Temperature (C)')
title('step change in Fa')
%
figure(2)
[yh11,xh11,th11]=step(numfrg(1,:),denn);
[yh12,xh12,th12]=step(numfa(1,:),denn);
[yh21,xh21,th21]=step(numfrg(2,:),denn);
[yh22,xh22,th22]=step(numfa(2,:),denn);
subplot(2,2,1)
plot(th11,yh11,'k-')
xlabel('Time(sec)')
ylabel('Reactor Temperature (C)')
title('step change in Frg')
%
subplot(2,2,2)
```

```matlab
plot(th12,yh12,'k-')
xlabel('Time(sec)')
ylabel('Reactor Temperature (C)')
title('step change in Fa')
%
subplot(2,2,3)
plot(th21,yh21,'k-')
xlabel('Time(sec)')
ylabel('Regenerator Temperature (C)')
title('step change in Frg')
%
subplot(2,2,4)
plot(th22,yh22,'k-')
xlabel('Time(sec)')
ylabel('Regenerator Temperature (C)')
title('step change in Fa')
numrc=numfa(1,:)+numfrg(1,:)
numrg=numfa(2,:)+numfrg(2,:)
Hrc=tf(numrc,denn)
Hrg=tf(numrg,denn)
%
[yrc,xrc,trc]=step(numrc,denn);
nprc=[2.4734];
dprc=[22  1];
td=2;
[yprc,xprc,tprc]=step(nprc,dprc);
xlabel('Time(sec)')
ylabel('Reactor Temperature (C)')
%
[yrg,xrg,trg]=step(numrg,denn);
```

```
nprg=[2.569];
dprg=[23 1];
td=3;
[yprg,xprg,tprg]=step(nprg,dprg);
%
figure(3)
subplot(2,2,1),plot(trc,yrc,'k-')
hold on
nprc=[2.4734];
dprc=[22 1];
td=2;
[yprc,xprc,tprc]=step(nprc,dprc);
plot(tprc+td,yprc,'k--')
xlabel('Time(sec)')
ylabel('Reactor Temperature (C)')
legend('Response','Approximation',4)
subplot(2,2,2),plot(trg,yrg,'k-')
hold on
nprg=[2.569];
dprg=[23 1];
td=3;
[yprg,xprg,tprg]=step(nprg,dprg);
plot(tprg+td,yprg,'k--')
xlabel('Time(sec)')
ylabel('Regenerator Temperature (C)')
legend('Response','Approximation',4)
```

5.2 Simulink programs

Simulink, a software package, depends on icons inset of functions to build simulation, solve mathematical equations and displaying plots. It contains many toolboxes for applications in control system such as PID feedback, Fuzzy logic neural network, optimization, curve fitting and ... etc. all these icons are representing in Simulink library browser where it contains relating to this work,

✓ Commonly used block: contains constant, sum, gain and mux.........etc.

✓ Continuous block: contains derivative, integrator state space and transfer function.........etc.

✓ Math operation block: contains abs, sum, sum of elements and math function, divide.........etc.

✓ Sink block: contains scope and display.........etc.

✓ Source block: contains step and ramp.........etc.

✓ Fuzzy Logic Toolbox.

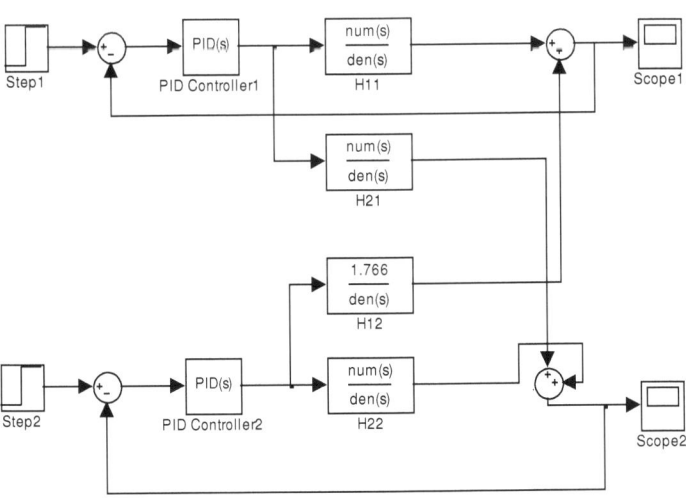

Figure (5.1). Block diagram for Interaction system using Simulink.

44

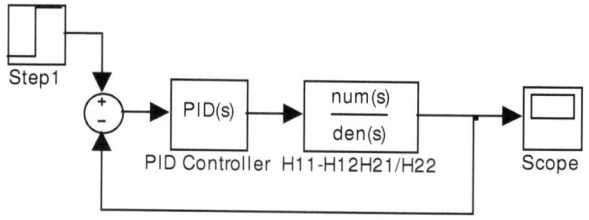

Figure (5.2): Block diagram for decoupling system using Simulink.

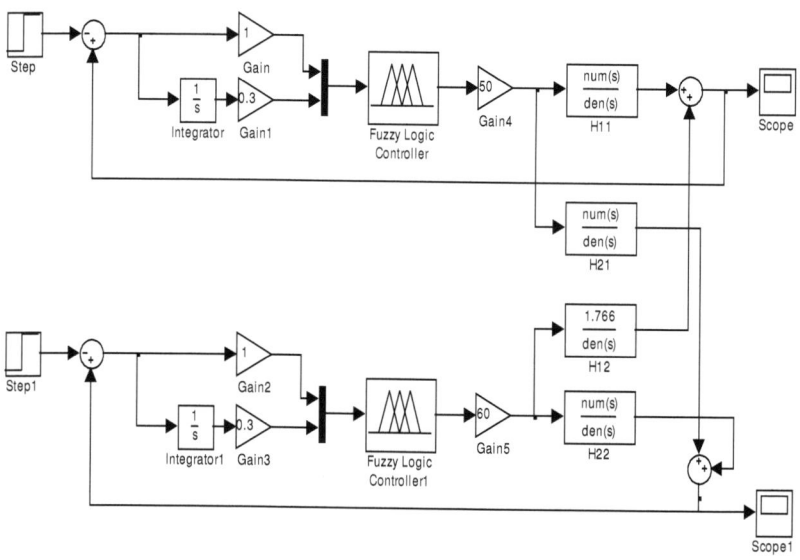

Figure (5.3): Block diagram of Fuzzy logic controller for Interaction using Simulink.

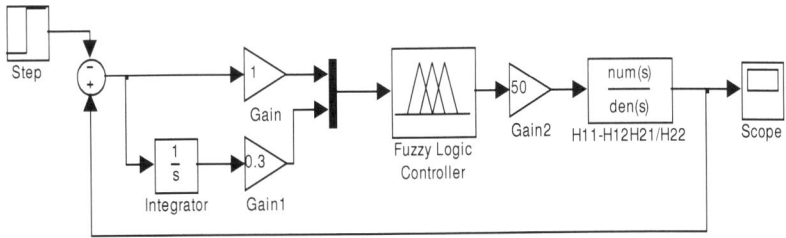

Figure (5.4): Block diagram of Fuzzy logic controller for decoupling system using Simulink.

The following diagrams show applying the Fuzzy logic control in the MATLAB program. Figure (5.5) shows the main window for Fuzzy logic control and the main data such as the error, change of error and the action should be identified. Figure (5.6) shows the window that concern with the error; it displays the range of error and the range of NB, NS, Z, PS and PB. Figure (5.7) shows the window that concern with the change of error; it displays the range of change of error and the range of NB, NS, Z, PS and PB of it. Figure (5.8) shows the window that concern with the action; it displays the range of action and the range of NB, NS, Z, PS and PB. Figure (5.9 a, b & c) shows the window that concern with the 25 rules.

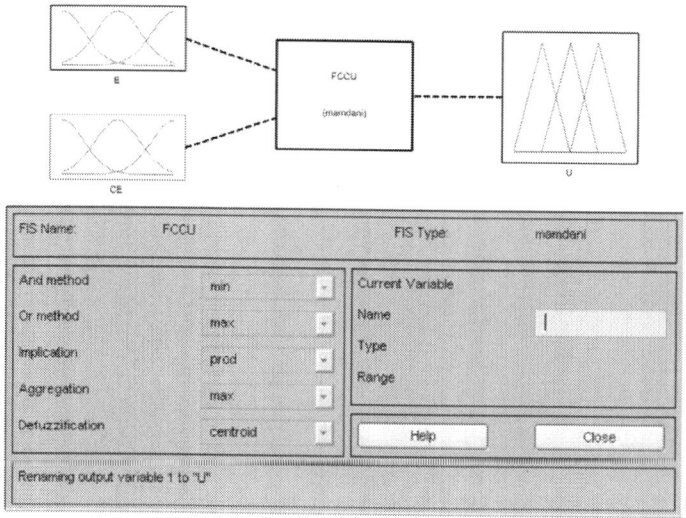

Figure (5.5) Main Diagram of Fuzzy logic controller.

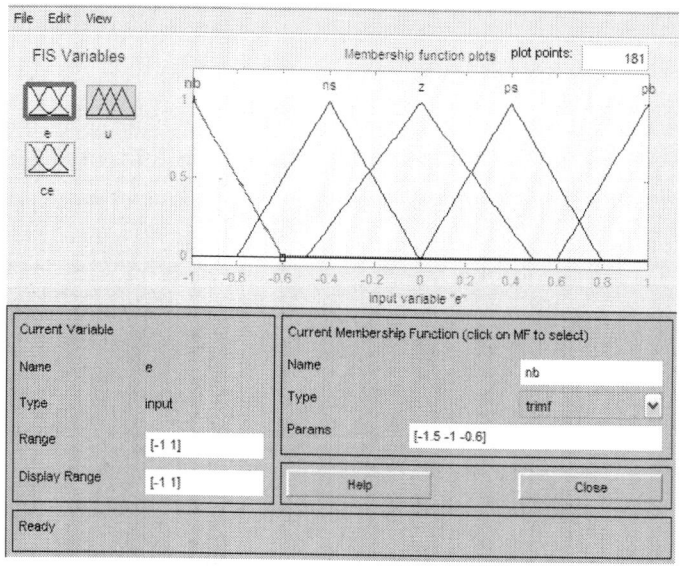

Figure (5.6) Diagram of error.

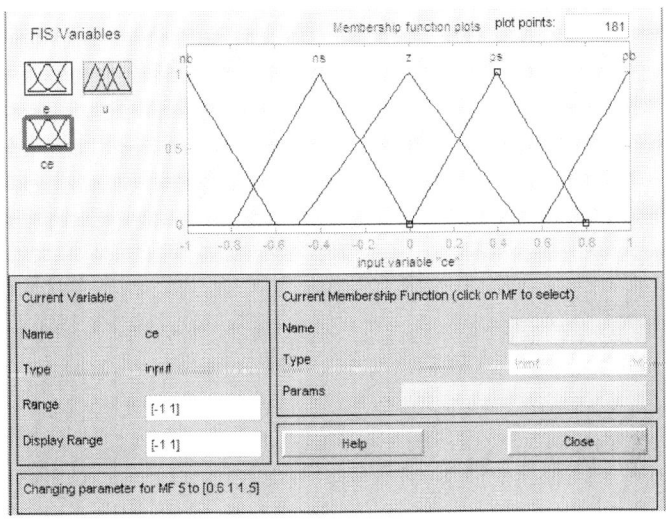

Figure (5.7) Diagram of change of error.

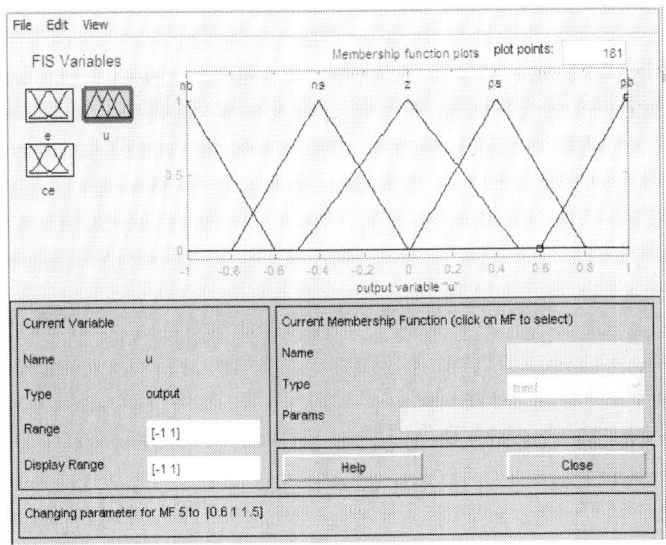

Figure (5.8) Diagram of action.

48

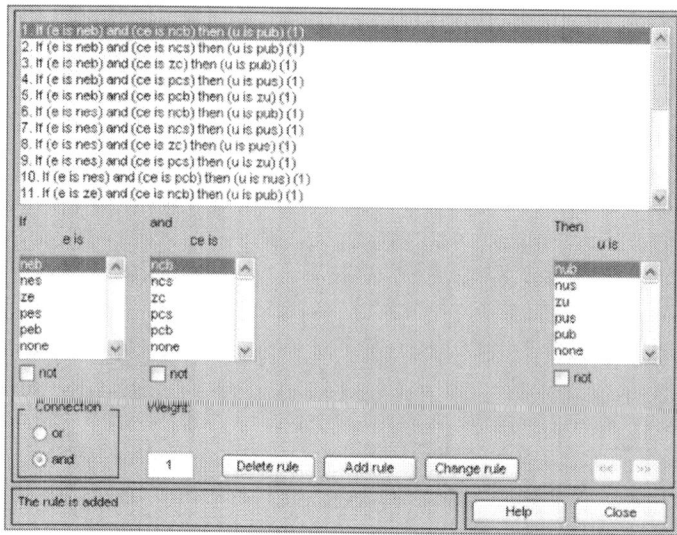

Figure (5.9a) Diagram of Mamdani.

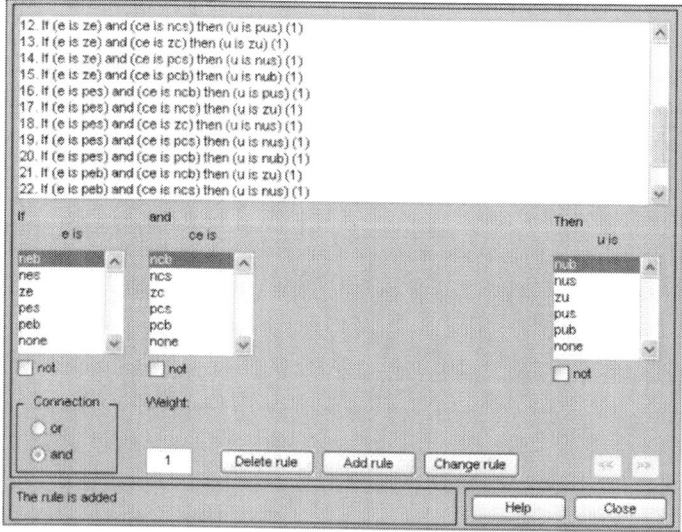

Figure (5.9b) Diagram of Mamdani.

49

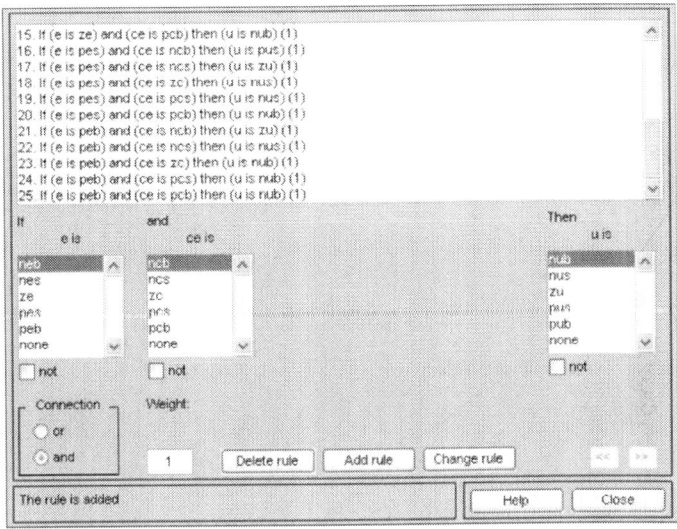

Figure (5.9c) Diagram of Mamdani.

References

Ali M.M., (1993), Rule Based Computer Control of Industrial Process using Fuzzy Logic, M.SC. Thesis, University of Technology, School of Control and Computers Engineering.

Arnon Arbel, Irven H. Rinard, and Reuel Shinnar, (1996), Dynamics and control of fluidized catalytic crackers. 3. Designing the control system: Choice of manipulated and measured variables for partial control, Ind. Eng. Chem. Res.,35 (7), 2215–2233.

Bollas G.M., Vasalos I.A., Lappas A.A., Iatridis D.K., Voutetakis S.S. and Papadopoulou S.A.,(2007), Integrated FCC riser—regenerator dynamics studied in a fluid catalytic cracking pilot plant, Chemical Engineering Science, Volume 62, Issue 7, Pages 1887-1904.

Brain Roffel and Ben Betlem, (2006), Process Dynamics and Control, modeling for control and prediction, John Wiley and Sons Ltd.

Donald R. Coughanowr, (1991), Process systems analysis and control, second edition, McGraw-Hill, Inc.

Gary J. H., Handwerk G. E. and Glenn E. Handwerk,(2001), Petroleum Refining Technology and Economics, Four Edition, Marcel Dekker, Inc.

Grosdidier, P., Froisy, B., & Hammann, M. , (1988), Proceedings of the IFAC workshop on Model based Process Control, Oxford: Pergamon Press, Pages 31-36.
Hobson G. D., (1973), Modern Petroleum Technology, Fourth Edition, Essex: Applied Science.

Jan Jantzen, (2007), Foundations of Fuzzy logic control, John Wiley & sons Ltd.

Kevin M. Passio and Stephen Yorkovich, (1998), Fuzzy Control, Addison Wesley Longman, Inc.

Lee E. and Groves F. R., (1985), Mathematical Model of the Fluidized Bed Catalytic Cracking Plant. Trans. Soc. Comput. Simulation, Pages 219-236.

Lewis W.K. and Gilliland E. R.,(2005), Massachusetts Institute of Technology, FCC History, PCK, Raffineria GmbH, Germany.

Luyben, W.L. and Luyben M.L.,(1997), Essentials Process Control, McGraw-Hall.
Luyben, W.L., (2007), CHEMICAL Reactor Design and Control, John Wiley & Sons, Inc.

McFarlane R. C., Reinman R. C., Bartee J. F. and Georgakis C., (1993), Dynamic Simulator for a Model IV Fluid Catalytic Cracking Unit, Comp. & Chem. Eng.., Pages 275-300.

Meyers R. A., and Hunt D. A., (1996), Handbook of Petroleum Refining Processes, Second Edition, McGraw-Hill.
Moro L. F. L. and Odloak D., (1995), Constrained Multivariable Control of Fluidized Catalytic Cracking Converters, proc. Cont.

Reza Sadeghbeigi, (2000), Fluid Catalytic Cracking Handbook. Design, operation and troubleshooting of FCC facilities, Second Edition, Gulf Publishing Company.

Shaymaa H. Khazaal, (2005), Computer Aided Design of a Fluidized Catalytic Cracking Unit using Iraqi feed stocks, MSC thesis, University of Technology, 2005.

MATLAB, version 7.10 (R2010a), the language of technical computing, www.mathworks.com.

Stephanopoulos, G., (1984), Chemical Process Control: an Introduction to Theory and Practice, Prentice-Hall India.

Uygun Ö ., Taşkin H., Kubat C., and Arslankaya S., (2006) Fuzzy FCC: Fuzzy logic control of a fluid catalytic cracking unit (FCCU) to improve dynamic performance, Computers & Chemical Engineering, Volume 30, issue 5, pages 850–863.

Wooyoung Lee and Alan M. Kugelmanl, (1996), Number of Steady-state Operating Points and Local Stability of Open-Loop Fluidized Catalytic Cracking Cracker, Ind. Eng. Chem. Process.

Zadeh L.A., (1989), Fuzzy relation Equations and Applications to Knowledge Engineering, Kluwer Academic Publishers, Holland.

Appendix: Data

Temperatures in (^{o}C)at Steady state	
Reference	25.0
Feed	200.0
Reactor	503.0
Regenerated Catalyst	720.0
Regenerator	715.5
Spent Catalyst	500.0
Steam	250.0
Mass flow rates in (Kg/sec) at Steady state	
Air	66.41
Coke	8.58
Flue Gases	75
Product	62.95
Regenerated Catalyst	454.79
Spent Catalyst	463.37
Mass in (Kg) at Steady state	
Flue Gases	750
Product	314.76
Regenerated Catalyst	4547.93
Spent Catalyst	2316.86
Specific Heat in (Kj/ Kg ^{o}C) at Steady state	
Air	1.08
Feed	2.268
Flue Gases	1.18
Product	2.687
Regenerated Catalyst	0.595
Spent Catalyst	0.583
Residence Times in sec at Steady state	
Reactor	5
Regenerator	10

2901103R00038

Printed in Great Britain
by Amazon.co.uk, Ltd.,
Marston Gate.